Let's Read About Food

Fats and Sweets

by Cynthia Klingel and Robert B. Noyed
photographs by Gregg Andersen

Reading consultant: Cecilia Minden-Cupp, Ph.D.,
Adjunct Professor, College of Continuing and Professional Studies, University of Virginia

WeeklyReader.
EARLY LEARNING LIBRARY

**For a free color catalog describing
Weekly Reader® Early Learning Library's
list of high-quality books, call 1-800-542-2595
or fax your request to (414) 332-3567.**

Library of Congress Cataloging-in-Publication Data available
upon request from publisher. Fax (414) 336-0157 for the
attention of the Publishing Records Department.

ISBN 0-8368-3056-3 (lib. bdg.)
ISBN 0-8368-3145-4 (softcover)

This edition first published in 2002 by
Weekly Reader® Early Learning Library
330 West Olive Street, Suite 100
Milwaukee, WI 53212 USA

An Editorial Directions book
Editors: E. Russell Primm and Emily Dolbear
Art direction, design, and page production: The Design Lab
Photographer: Gregg Andersen
Weekly Reader® Early Learning Library art direction: Tammy Gruenewald
Weekly Reader® Early Learning Library production: Susan Ashley

Note to Educators and Parents

As a Reading Specialist I know that books for young children should engage their interest, impart useful information, and motivate them to want to learn more.

Let's Read About Food is a new series of books designed to help children understand the value of good nutrition and eating to stay healthy.

A young child's active mind is engaged by the carefully chosen subjects. The imaginative text works to build young vocabularies. The short, repetitive sentences help children stay focused as they develop their own relationship with reading. The bright, colorful photographs of children enjoying good nutrition habits complement the text with their simplicity and both entertain and encourage young children to want to learn — and read — more.

These books are designed to be used by adults as "read-to" books to share with children to encourage early literacy in the home, school, and library. They are also suitable for more advanced young readers to enjoy on their own.

— *Cecilia Minden-Cupp, Ph.D.,*
Adjunct Professor, College of Continuing and
Professional Studies, University of Virginia

I like to eat fats and sweets. They taste good.

We choose from six different kinds of food. We need to eat all six kinds every day to stay healthy.

fats and sweets

milk and cheese

meat

vegetables

fruit

bread and cereal

Ice cream and butter have fat in them. Candy, cake, and cookies taste sweet.

Too many fats and sweets are not good for my tummy.

Too many fats and sweets are also not good for my teeth.

My body does not get many vitamins from these foods.

I try not to eat many fats and sweets. That is one good way to stay healthy.

Dad says having one treat a day makes it special.

I am hungry. Should I eat a cookie or a piece of fruit?

Glossary

fats–oily substances found in plant and animal tissue

sweets–foods made with lots of sugar

vitamin–one of the substances in food that is needed for good health

For More Information

Fiction Books

Goldstein, Bobbye, and Kathy Couri. *Sweets & Treats: Dessert Poems.* New York: Disney Press, 1998.

Seuss, Dr. *The Butter Battle Book.* New York: Random House, 1984.

Nonfiction Books

Christian, Eleanor, and Lyzz Roth-Singer. *Let's Make Butter.* Mankato, Minn.: Pebble Books, 2000.

Frost, Helen, and Gail Saunders-Smith. *Fats, Oils, and Sweets.* Mankato, Minn.: Pebble Books, 2000.

Web Sites

Candy USA
www.kidscandy.org/
For fun facts about candy

Name That Candy Bar
www.sci.mus.mn.us/sln/tf/c/crosssection/namethatbar.html
For a quiz that tests your candy bar knowledge

Index

About the Authors

Cynthia Klingel has worked as a high school English teacher and an elementary school teacher. She is currently the curriculum director for a Minnesota school district. Cynthia Klingel lives with her family in Mankato, Minnesota.

Robert B. Noyed started his career as a newspaper reporter. Since then, he has worked in school communications and public relations at the state and national level. Robert B. Noyed lives with his family in Brooklyn Center, Minnesota.